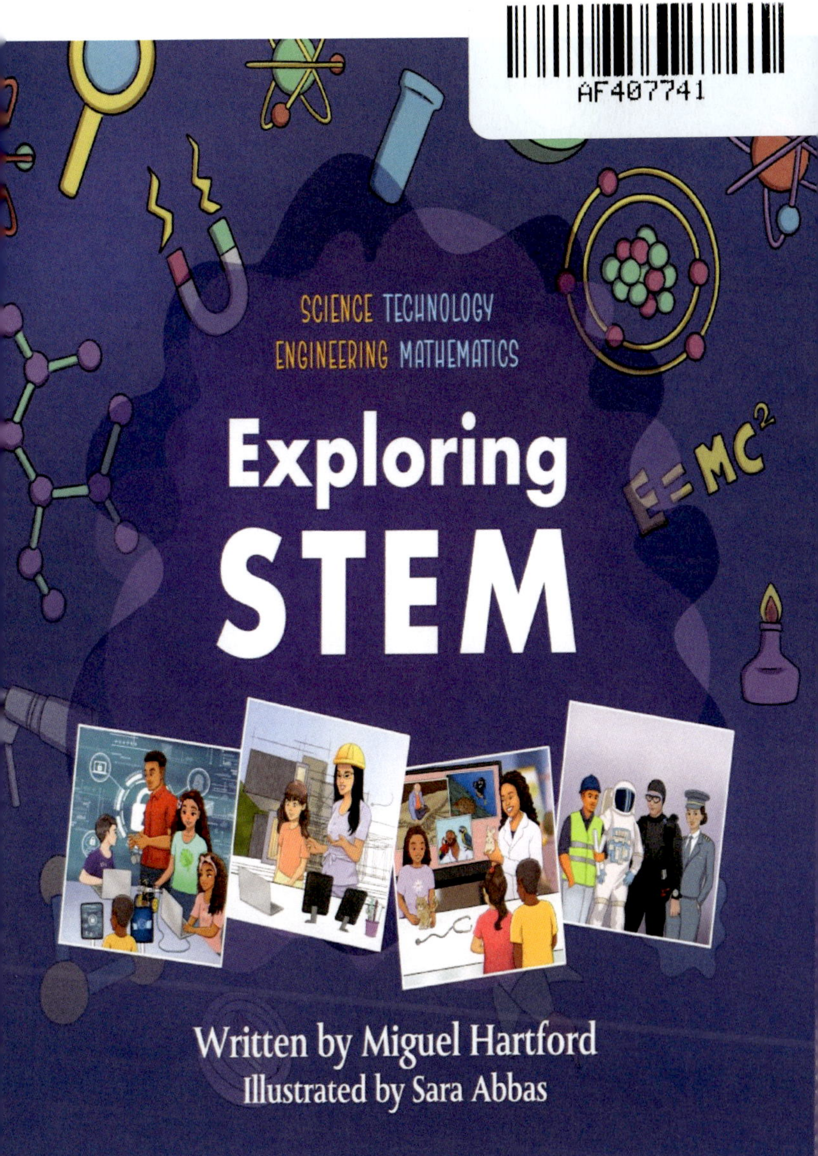

SCIENCE TECHNOLOGY ENGINEERING MATHEMATICS
Exploring STEM
$E=MC^2$
Written by Miguel Hartford
Illustrated by Sara Abbas

Table of Contents

EXPLORING STEM

Written

by

Miguel Hartford

What is STEM and Why is it important

STEM is Science, Technology, Engineering and Mathematics. It is important, because STEM jobs are some of the most demanding jobs today. This might not change in a long time, if at all. As technology advances, there is a higher demand for STEM workers. The more trained and educated a STEM worker is, the more likely they will have a job.

Computer Science is the highest paid major. It is also the least regretted major. This is for multiple reasons such as salary, job security, development and continuous learning. The salary or the amount of money people make, is much more than those of other careers. STEM fields provide job security, as the STEM jobs are in high demand, and there are many positions available. STEM jobs keep innovating or changing. STEM requires continuous learning, as they keep training and develop their education, as technology keeps changing. This is a good thing. The people who work these careers, and have this knowledge, will always be needed.

Examples of STEM careers

Doctors, Registered Nurses, Pharmacists, dentists, orthodontists, Chemical Engineers, Civil Engineers, Architects, Computer Programmers, Software Engineers, illustrators, graphic designers, pilots and astronauts are some common STEM careers. Other STEM careers that are not as common are radiologists, x-ray technicians, radiologist technician, data analysts, data scientists, scrum masters, and business analysts.

College and STEM Majors

Some STEM majors are Biology, Chemistry, Computer science, Engineering, Earth sciences, Health Sciences, Astronomy, Information Technology,

Mathematics, Physics, and Nursing. The first level of college degrees is an Associate's Degree. This is earned a 2-year college or vocational school, and some 4-year Colleges.

Most people with STEM careers receive a Bachelor of Science. A Bachelor's Degree (BS) is a degree that someone earns from a 4-year college or university after receiving a high school diploma or equivalent certificate. The next level degree is a graduate degree. This is a Master of Arts (MA), Master of Science (MS), or Master of Fine Arts (MBA). This usually takes another 2 years. Then, there is a Doctorate or Professional Degree for professions like medicine and law. It is mandatory for some careers like doctors, nurses, pharmacists and engineers to get a bachelor's degree and to continue their higher education to earn a master's degree and for some a doctorate degree. Doctors earn an M.D., dentists earn their D.D.S., pharmacists earn a Pharm. D, nurses earn a D.N.P. scientists can earn a D.S.C., and engineers can earn a D.E.S. These are some examples of doctoral degrees.
 Not everyone who works in STEM needs a college degree. Some careers can begin with certifications at community colleges or vocational schools, as well as personal online training. Some STEM jobs that do not require a degree are Vetinary Assistant, Nursing Assistant, Pharmacy Technician, Cybersecurity roles, Computer User Support Technician, Ultrasound Technician, Graphic Designer, Dental Lab Technician, Web Developer, Dental Hygienist and others.

It's never too early to learn about STEM

Students should start learning about STEM as young as possible. This begins with questions, reading and projects to learn about science. No one is too young to learn STEM. A lot of middle schools offer classes to teach about coding and robotics.

Coding and Robotics

There are STEM games, programs and camps for kids. Two important parts of STEM to learn are coding and robotics. Coding is the process of creating instructions for computers using programming languages. This is how we tell the computers what to do, to control the computers and electronics we use. There are free games to teach children how to code. This gives them practice for when they get older and get into coding and other levels of

STEM. Coding and robotics allow students to demonstrate critical thinking and problem-solving skills.

Robotics, or the design, construction, operation and use of robots is also important for kids to learn. This can be fun and useful. Robotics controls so much of the world today, as well as technology of the future. It is important that students interested in STEM get involved in robotics at school and/or at home and in other programs to learn, advance and showcase their careers.

Middle school students are building robotics projects. This ranges from potatoes and dirt batteries to budling gears and armed robots that grab objects and robotic cars. Students can build robotics projects at any age. A teenager named Sergio Peralta was born without his right hand. In January 2023, his classmates in an engineering class built him a robotic hand after looking at designs online. This gave him use of his right hand for the first time.
Artificial intelligence (AI), is the theory and development of computer systems able to perform tasks that normally require human intelligence such as sight and speech. It is a common term among robotics today. This has affected and improved our daily lives from everyday benefits, such as virtual agents to organize meetings to helping with problems and shopping. It has evolved to other areas. This can be machine learning such as machines used to provide safety in the workplace. There is also healthcare, and advancements in the medical field such as analyzing

patient data, to utilizing equipment for surgery, that is now more advanced than ever.

STEM Careers

Medical Careers

Students are familiar with some STEM careers, but not all of them. When it comes to the medical students have all visited a general medicine doctor, who can practice and diagnose common symptoms or the dentist who keeps people's teeth and gums healthy with checkups and cleaning. A pediatrician is a doctor who specializes in children's medicine. Not everyone has been to an orthodontist. This is a dentist who treats irregularities in the teeth and jaws. An orthodontist is the dentist who will put braces, aligners or retainers on someone's teeth to make them straight.

Nursing is also a familiar profession to students, but there are different kinds of nurses. An LPN is a Licensed Practical Nurse. This profession requires school for 2 years. They work under Registered Nurses and check patients' vital signs and basic health conditions, as well as administrative work. Registered Nurses or RNs are very common, and have more duties than LPNs. This is a high paying career and requires a bachelor's degree. RNs work with doctors, such as arrange patient care and educate patients on their health, as well as giving support to patients and their families through education. Cardiac

Nurses are in demand due to the high number of patients dying from heart disease. Certified Registered Nurse Anesthetist is one of the highest paying nursing jobs. These nurses assist doctors with administering anesthesia to patients during surgeries. Anesthesia is a medical treatment that prevents patients from feeling pain during procedures like surgery, certain screening and diagnostic tests, tissue sample removal and dental work,

RNs and other nurses can increase their education and training to become Nurse Practitioners. Nurse Practitioners can give primary and secondary care to patients as they can give physical examinations, order tests and provide medications. ER Nurses work in the Emergency Room and provide immediate assistance to patients, as well as evaluate and make sure patients are stable. Travel Nurses can travel to different cities. Nurse Midwives specialize in taking care of pregnant women. Oncology Nurses help patients with cancer. Mental Health Nurses have a specialty of diagnosing and treating psychological disorders in patients. Surgical Nurses assist with patients and provide care in the operating room before and after surgery. Public Health Nurses provide outreach to the public, as they teach about health education. There are many other nursing careers to choose from.

The doctors and surgeons

Although students are familiar with doctors in general, there are numerous medical doctors. General Medicine or Family Medicine doctors treat common medical conditions. They can then refer patients to a specialist, meaning someone who practices medicine in a particular area.

OB-GYNS practice Obstetrics and Gynecology. They specialize in women's health and deliver babies when they are born. Pediatricians are doctors that treat children. Cardiologists study and treat heart diseases and heart abnormalities. Optometrists treat eye diseases and vision problems. Anesthesiologists give patients anesthesia, a medical treatment that prevents patients from feeling pain during procedures like surgery. They evaluate, monitor, and supervise a patient during surgery as they make sure their breathing, blood pressure, body temperature and other basic conditions are safe. Veterinarians are doctors that provide medical attention to animals.

There are also different types of surgeons. General surgeons will focus on the abdomen, as well as head and neck. Other surgeons specialize in certain areas. Orthopedic surgeons treat problems of the body's bones muscles and joints. Plastic surgeons repair, replace and reconstruct physical defects of the skin or other areas of the body. They may perform surgery due to birth defects, injuries, illnesses or other reasons.

Other Medical Careers

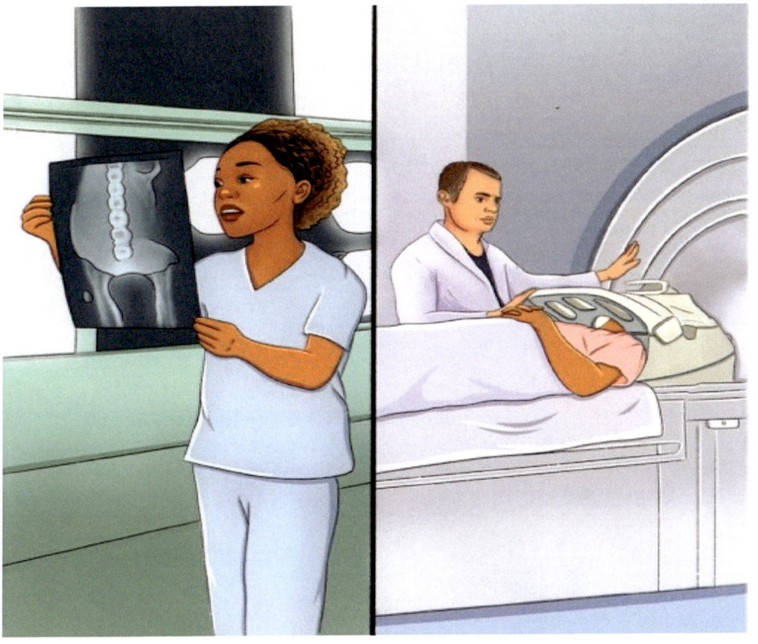

Not everyone in the medical field is a physician or nurse. Pharmacists dispense medicine to patients prescribed by doctors and provide safe instructions for taking the medicine. X- ray Technicians take x-rays of patients. Radiologists interpret the medical images such as X-rays and MRIs (Magnetic Resonance Imaging). This is when a patient is goes into a machine to detect brain tumors, traumatic brain injury, or diagnose the effects of strokes dementia, infection, causes of headache and other medical injuries. They then diagnose patients and recommend treatment. Dieticians and nutritionists teach people about healthy eating and safe medication. Chiropractors help to treat patients through spinal adjustments and manipulation. Physical Therapists help patients improve their range of motion due to an illness or injury. Dental hygienists clean patients teeth and check for signs of decay and gum diseases. Speech therapists assess, diagnose and treat communication and swallowing disorders. Their patients range from victims of strokes, brain injury, developmental delays and autism.

Cheers to the Architects and Engineers

Engineers are some of the most common STEM careers. They innovate or change the world we live in. There are Civil Engineers, Mechanical Engineers, Aerospace Engineers, Chemical Engineers and more. Architects and civil engineers often work together with planning and designing, but there is a difference between the two. Architects create as they design the structures of homes, buildings, or other projects such as railroads, bridges, canals and more. Civil Engineers plan, construct and improve the infrastructure, as they focus on safety. It is important that students wanting to get into these careers focus on math and science on advanced levels. These careers require a Bachelor's Degree.

Mechanical Engineers design machines powered by electricity, gas, and steam power to problem solve. They can design machines inside buildings like elevators and escalators to transmissions, aircrafts and control systems. Chemical Engineers develop and design chemical manufacturing processes. This career involves combining chemistry, math, biology and physics to solve problems. They work with products such as fuel, drugs, food and other products. Industrial engineers use math and science to solve complex processes to develop and improve systems.

Careers in the Air and Sky

STEM careers such as pilots who can fly planes or aircraft and astronauts, who can fly into space. Pilots usually need a bachelor's degree in transportation, business or engineering. Astronauts must have a Master's Degree in a STEM field. There are also aerospace engineers who develop aircraft and spacecraft. Astronomers observe objects in the universe, such as planets, stars and galaxies. Computer hardware engineers test computer systems and equipment that measures activity in outer space or on Earth. Electronics Engineers make sure the equipment is safe. Mechanical Engineers make sure sensors, tools engines and other machines are successful to support missions to space. There are different types of technicians to assist the engineers in areas like testing and results.

There are also STEM careers on the water. A common career is a Marine Biologist, who researches migration and evolution patterns of marine life, the health of marine environments, the human impact on marine animals and reefs, and how to repair damage to ecosystems. Ecologists study the relationships between animals, plants and the environment. Aquatic veterinarians give medical care to animals that live in the water. Marine Engineers design marine vessels like cargo ships and oil rigs, as well as repair engines, pumps and other technical equipment. Marine Archeologists study history while scuba diving.

Sciences

There are many STEM careers under the branch of science. Meteorologists are physical scientists who study the weather through observation, studying and forecasting. Archaeologists study human history and prehistory through the excavation of sites and the analysis of artifacts and other physical remains. Paleologists study fossil animals and plants. Zoologists and wildlife biologists study animals in other wildlife and how they interact with their ecosystems. Geologists study the composition, structure and other physical attributes of the earth. Physicists plan and conduct scientific experiments and studies to test theories and to discover properties of matter and energy. Biotechnologists create and improve products and processes for agriculture, medicine, and conservation using biological organisms. Environmental scientists use their knowledge of the natural sciences to protect the environment.

Computer Careers

There are numerous computer related STEM careers. Computer Science is one of the most diverse majors and can result in a career in so many areas. This is why it is one of the highest paid majors and students who major in it are very happy with their choice. Information Technology or IT, is the use of computers to create, process, store, retrieve and exchange all kinds of data and information. This ranges from repair and protecting data on computer systems to developing and supporting. Cyber Security is protecting computer systems from unauthorized access or being damages or made in accessible.

Design is a large area of computer careers. Software engineers or software developers create, design, develop and maintain, test and evaluate computer software. Illustrators can draw, paint or using other tools to create art. This can be seen in books, magazines, movies and TV. Graphic designers create visual concepts by using computer software or by hand to create visual on products we use, as well as forms of media.

Mathematics and Business

There are many careers that focus math and business. A career with math does not sound fun to many students, but some people love it. STEM math careers can be math teachers and professors. There are also careers such as analysts and scrum masters. Finance and statistics are STEM majors. Data analysts collect key information for businesses and use that to solve problems. Business analysts improve processes and recommend changes for companies with the data and information. A career term not most people aren't familiar with is a scrum master. Scrum masters are the facilitators for agile development teams.

Colleges to Attend

There is no particular college to go to as a STEM major. Students should choose whichever school fits them best. Most 4-year colleges or universities will offer common majors such as nursing, computer science, biology and some engineering programs. Some school specializes in certain academic programs. Different schools offer different forms of support for students to succeed. There are internships, where students get to be trainees for companies without pay. There are recruiters and job fairs where companies come to colleges and universities to hire students. Another way schools can support students is with culture, such as groups for women and minorities.

Most popular colleges for STEM

Some universities are specifically designed for Science and Technology, such as; MIT (Massachusetts Institute of Technology), Georgia Institute of Technology, California Institute of Technology, Stevens Institute of Technology, Missouri University of Science and Technology, and Michigan Technological University.

Other popular colleges for STEM are, Harvey Mudd College, Stanford University, University of California, Berkley, University of California, San Diego, Texas A&M University, University of Illinois-Urbana Champaign, University of Michigan, Rose-Hulman Institute, and Rice University.

College can be expensive. Some students who cannot afford to pay for college receive student loans. This means a financial company pays the school for the student's tuition, and the student pays them back over the years. However, another great route to take for STEM career is through the military. The United States Coast Guard Academy, The United States Air Force Academy and the United State Naval Academy have great STEM program. Those attending will serve in the United Stated States military branch. This results in the reward of the military paying for their education.

HBCUs and STEM

Some schools have STEM programs that stand out. Some also work with high schools, to get STEM students early. HBCU's or Historical Black Colleges and Universities are a major factor for African American students having successful STEM careers. African American students from lower income communities tend to have better success at HBCUS, as they tend to offer specialized help for students who need to develop their skills that other universities don't have. Some top HBCUs for STEM are Alabama A&M University, Alabama State University, Ala Hampton university, Florida A&M University, Howard University, Prairie View A&M University, North Carolina A&T University, Morgan State University, Jackson State University and Spelman College.

Students who earn STEM often stand out among their peers. For example, in the state of Louisiana, Xavier University and Dillard University graduate the most African American students who will become doctors and pharmacists. These students have a higher salary than their African American peers in the state. NCAT or North Carolina came in number one among HBCUs for the most HBCU graduates with 42.8 of graduates completing their STEM program in 6 years. Howard University, is a large HBCU with a program called Karsh STEM Scholars, which helps high school students use their classes towards their STEM college program.

Location

Where a school is located can benefit what the university can offer. A good example of this is NASA's Marshall Space Flight Center in Huntsville, Alabama partnering with Alabama A&M University for programs to have high school students for their Summer High School Apprenticeship Research Program and hire graduates from this HBCU. This helps students develop their STEM careers early, and with advanced tools and resources.

Location also affects how much a person's salary will be. Salary, is the amount of money of person makes at their job. States like California, Massachusetts, New York, New Jersey, Maryland, Connecticut, Alaska and Washington DC pay higher salaries. However, these states also have higher costs of living, meaning it costs more to live there. States that have the best high salaries and low cost of living are; Texas, Utah, Minnesota, North Carolina, Iowa and Virginia. Many graduates and other professionals tend to move to these states so they can make more money and pay less to live in areas like groceries, housing, transportation, education healthcare, and more.

Benefits of a STEM career

There are great benefits of STEM careers. One of the main benefits is innovation. Technology is constantly changing. Workers in STEM are required to learn to new ways to work and stay updated on education in their work fields. This makes STEM workers highly qualified.

Some include travel. Many jobs like data analysts, scrum masters, software engineers, programmers and cybersecurity can be fully remote, meaning a job to work from home, or hybrid where some work days are in office and the other work days are at home. Travel can also be a benefit. Some STEM careers require or allow travel. Relocation is also an option, as STEM careers can allow workers to live almost anywhere.

STEM careers pay the Highest salaries The average salary in the United States is $54,132.

In almost every state, the highest paying job is a STEM career. This includes surgeons, anesthesiologists, dentists, family and general practitioners, internists, OBGYNs, Oral surgeons, orthodontists, pediatricians, physicians and podiatrists.

Many STEM salaries pay much more than the average salary. A radiologist's salary ranges from $385,000 to $513,000. The Median salary for Registered Nurses is $77,000. Nurse Practitioners average $111,000. Anesthesiologists' salaries range from $360,000 to $471,000. Civil Engineers salary ranges from $53,000 to $110,000. Chemical Engineers salary ranges from $63,000 to $150,000. Data Analysts salary ranges from $48,000 to $110,000. Scrum Masters salary ranges from $$58,000 to $120,000.

People Needed in STEM Careers

Diversity is still needed among STEM Careers. Asian and White Students are the highest population among STEM graduates. Black and Latino students make up a small minority. There are programs geared towards decreasing this gap. Women make up only 25% of computer careers. They are also 74% of healthcare practitioners and technicians and 38% of surgeons. Women lead the vetinary field at 64%. Black and Latino women each make up only 2% in STEM careers. It is necessary that these numbers increase.

Why Choose a STEM Career

Students should choose a STEM career as STEM jobs are permanent in society. Overall, STEM students make higher salaries than most people. Some STEM professionals make the highest salaries in the country. STEM careers have a wide variety to choose from, so anyone interested in STEM can find a career that suits them.

What to do now, for a successful STEM Career Later

It is never too early to plan for the future. Students who make good grades in elementary school, are able to get into gifted and magnet schools in middle school. Middle school students can take robotics classes, boot camps and play coding games. This will give them an early start to advance their skills. The earlier something is learned, the more practice they will have, and the better they will become at it.

There are some classes that middle school students can take that will transfer to high school. This gives them an advantage, as they are able to take higher level math classes. Students who take higher level math classes in high school can use some of these courses that transfer as an equivalent for college, which means they don't have to take them there, and can earn their bachelor's degree in less time. They are also more competitive for college and have more skills more college as well as careers later.

It's Important that students take the classes they need for the STEM program that they are interested in. Higher math classes such as calculus and statistics. Computer science is also very important as it will be applied to most STEM careers. Robotics can be very beneficial to those interested in become a Computer Scientist or those interest in Engineering such as mechanical, software, and aerospace or other areas. Those interested in Nursing and become doctors should focus on math and science courses that apply such as chemistry, biology, physics and others. This will help students to develop skills and interests earlier, and this will transfer over to their careers.

Consider Choosing STEM

STEM is the present and future. Technology and way the world works is forever changing. Engineers are constantly creating things to make the world better, more advanced, easier and exciting. Students who choose STEM careers can look forward to a lifelong career. They can choose to stay in the same field, or switch to a different one if their interests change, as they progress in life. Choosing a STEM career can provide a high salary for financial security and those choosing to work in STEM careers can make a difference and be a part of changing technology and the world we live in.

ABOUT THE AUTHOR

Miguel Hartford grew up in Baton Rouge, Louisiana with her parents and two older brothers. She is a Southern University A & M College graduate with a Bachelor's of Science in Business Management. Her nieces, nephews and godson Caleb are her inspiration for the children's series. Growing up she was exposed to other races and ethnicities in school, but she later learned that this was not the case for everyone. She was introduced to different activities such as art, gymnastics, roller and ice-skating, and swimming regularly. Her goal is to expose students to the differences and likeness of others at an early age, and reading provides children exposure and teachable moments.

* 9 7 9 8 8 6 9 0 5 6 8 3 2 *